People of the Middle Ages

Blacksmith

Melinda Lilly

Original illustrations by Cheryl Goettemoeller

Rourke

Publishing LLC
Vero Beach, Florida 32964

www.rourkepublishing.com

For Wolfgang

PICTURE CREDITS: Page 5, detail of "Hartmann von Starkenberg," (Cod. Pal. Germ. 848, fol.256v) from the *Codex Manesse*, courtesy of the University of Heidelberg; Page 6, detail of MS. Top. Gloucs. d.2, fol. 24, courtesy of the Bodleian Library, University of Oxford; Page 9, detail of MS. Bodl. 614, fol.4, courtesy of the Bodleian Library, University of Oxford; Page 14, Photo by Scott M. Thompson, detail from *Weiss Kunig*; Page 18, detail of "Regenbogen," (Cod. Pal. Germ. 848, fol.381r) from the *Codex Manesse*, courtesy of the University of Heidelberg; Page 22, "Armourers at Work," (ROY. 16. G. V. f 11) by permission of The British Library; Page 26, The Metropolitan Museum of Art, The Cloisters Collection, Hewitt Fund, 1913. (13.138.1); Original art on page 25 is by Patti Rule. Cover illustration and original art on pages 10, 13, 17, 21, 29 by Cheryl Goettemoeller

Cover illustration: This blacksmith from the Middle Ages (years 500 to 1500) beats a red hot iron bar into the shape of a sword.

Editor: Frank Sloan

Cover design by Nicola Stratford

Library of Congress Cataloging-in-Publication Data

Lilly, Melinda
    Blacksmith / Melinda Lilly
      p. cm. — (People of the middle ages)
    Includes bibliographical references and index.
    Summary: Describes the life and work of a blacksmith living in Europe during the Middle Ages.
    ISBN 1-58952-226-5
      1. Blacksmiths—Europe—Juvenile literature. 2. Blacksmithing—Europe—History —To 1500—
Juvenile literature. [1. Blacksmithing—History. 2. Blacksmiths. Middle Ages.] I. Title.

TT220 .L55 2002   2001056512
682'.094'0902—dc21

Printed in the USA

CG/CG

Table of Contents

What's Your Name?

What's in a name? If your last name is Smith, your family may have been European blacksmiths of the Middle Ages. During this time (from the year 500 to 1500), people added last names. With five Marys and two Pauls in town, it was hard to know who was who!

People chose their last names for many reasons. John's son might have taken the name Johnson. A proud blacksmith would have chosen the name Smith.

A blacksmith of the 1300s works on a helmet as a lady brings him food.

A Very Important Person

What if you were a blacksmith in the year 1100? You would make tools from a black metal called iron. That is why you are called a blacksmith! Life would be harder without the metal tools you made.

For farmers, you would make horseshoes and plows. No one wants to try cooking without iron pots and spoons. You might make candlesticks and jewelry from gold. Without stirrups, knights would fall off their horses every time they swung their swords. What would people do without you?

A smith of the Middle Ages who worked with gold poses proudly with one of his candlesticks.

Red Hot

Clang, clang! The dawn bell rings as you enter your workshop. You light the charcoal in the furnace. Your son pumps the **bellows**, making wind to make the fire hot.

Phew, it's hot! You carefully tend the fire. Its color shows its temperature. When it's red, it's hot enough to soften iron. The iron needs to be soft before you can shape it.

A young man works the bellows to heat up a fire. This picture is from a handmade book of the Middle Ages.

Beating Iron

You heat an iron bar, and then pull it from the fire before it melts. Like a pair of drummers beating the same drum, you and your assistant beat the iron in rhythm. Beating it makes it stronger and thinner.

Next, you use tools to shape it. When it gets too cool to shape, you put it back in the fire. After your work is finished you dunk it in water. You've made a knife!

A blacksmith and his assistant beat iron.

Family Business

Your family helps you in your business. Your son cleans your tools by scrubbing them in a bucket of sand and vinegar. He sweeps the shop and works the bellows. As he becomes older, he will join you in hammering and shaping.

Your wife and daughter help you sell the things you've made. They make sure you get a fair price. Each week at the village market they sell pots, hammers, and other goods.

The blacksmith makes a horseshoe while his son cleans a tool and his daughter sells a spoon.

From Spoons to Swords

The job of a blacksmith changes during the Middle Ages. At first, smiths make everything from spoons to swords. As the years pass, many **specialize**. They craft only one type of product.

Want to work for knights and kings? Become an **armorer**. You can become famous. Some armorers are still remembered for the royal armor they made.

A smith (seated on left, in cap), in his workshop, with his assistant and Emperor Maximilian.

Apprentice

You want to be a blacksmith who makes armor. How do you become one? First, you train as an **apprentice**. You serve and learn from a master craftsman.

Before you begin, you must swear to help the blacksmith and his wife. In return, the blacksmith promises to teach you his trade. He also will feed and clothe you. At about age eight, you move in with him and his family.

An apprentice greets the blacksmith he will serve.

Follow the Rules!

Here are the rules for armorers' apprentices:

No swearing. No fighting with swords. No dancing. No playing tennis or bowling. No playing cards or dice. Stay away from bars and banquets.

Don't wear a shirt with a ruffle. The colors of **hosen** (long socks) that are allowed are white, brown, and blue. Your **breeches** (knee-length pants) must be plain.

If you are caught breaking a rule, you will be whipped in front of the other apprentices. Better mind your socks!

An apprentice (the boy on the left) bores a hole while blacksmiths discuss work. This picture is from a songbook made in Switzerland in the 1300s.

Joining the Club

You've been an apprentice for twelve years. One day, master craftsmen come to judge your work. They look closely at each link of the chain mail armor you made.

At last, they say your work is good. You're finally an armorer! You can join the **guild**. The armorer's guild is a professional club for master smiths. The masters of the guild lead you to a feast in your honor. **Hussa!**

An apprentice waits to hear the judgment of the master craftsmen.

Making Armor

You and the workers in your new shop make chain mail and plate armor. Both take months of work.

To make chain mail, first you must make wire. Next, bend the wire and cut it into rings. Link thousands of rings together in a tight weave to complete the chain mail.

You make plate armor from sheets of metal—often steel. Hammer the metal. Weld the thin metal sheets together. Shape each piece of armor to fit the knight. Decorate and make them shine!

An upper-class woman supervises men working on chain mail and plate armor. Another man plays the flute. This picture is from the 1400s.

23

Battling Empty Armor

You finish making a suit of armor. Now you have to test it. Set it up on a stand or hang it from a tree.

Attack! Jab the empty armor with a sword. Aim and shoot arrows at the metal enemy. Stab it with a spear. Yah!

If nothing pierces the metal, you've done your job. The knight who wears it will be well protected in battle. Time to sell your masterwork.

A blacksmith tests armor.

26

The Mark of Quality

Your guild may sell your armor in far-off countries. You claim your work by putting a mark on it. Your mark is like a signature. It shows everyone that you made the armor.

If you are famous for your skill, others may try to put your mark on their work. By pretending that you made it, they can charge a higher price for the armor.

An armorer's shop of the Middle Ages

Legends of the Blacksmith

How you do your job is a mystery to many people. Legends tell of spells chanted over swords. Dwarves get the credit for making armor you slaved over for months!

Smiths even have their own **saint**, or holy person of the Catholic Church. One day while Saint Eligius was working, the devil leaped out of his fire! Saint Eligius pinched the devil's nose with red-hot tongs. Yelping, the devil disappeared. Nobody tangles with a good blacksmith!

Saint Eligius pinches the devil's nose.

Dates to Remember

476	Last Roman emperor overthrown (Romulus Augustulus)
500	Beginning of the Middle Ages
700s	Stirrups arrive in Europe from China
1100–1200	Waterwheels first used to power bellows
1350	Blast furnaces used by Swedish blacksmiths
1320	Small cannon invented
1500	End of the Middle Ages

Glossary

apprentice (uh PREN tis) — a person who works for a master craftsperson in order to learn a trade

armorer (AR mur er) — a person who makes or repairs weapons or armor

bellows (BEL ohz) — a device for making a strong current of air

breeches(BRICH iz) — knee-length pants

guild (GILD) — a professional club of the Middle Ages that set standards and helped its members

hosen (HOE zun) — a stocking or long sock

hussa (huh ZAH) — Hooray, as said by people in the Middle Ages

saint (SAYNT) — a holy person recognized as such by the Catholic Church

specialize (SPESH uh lize) — To work in a specific field or on one subject

31

Index

Further Reading

Bachini, Andrea. *The Middle Ages.* Barrons Juvenile, 1999.

Dawson, Imogen. *Clothes and Crafts in the Middle Ages*. Gareth Stevens, 2000.

Jordan, William C. (editor). *The Middle Ages: A Watts Guide for Children.* Franklin Watts, Inc. 2000.

Websites to Visit

Time Line of Inventions of the Middle Ages
www.scholar.chem.nyu.edu/~tekpages/Timeline.html
Life in the Middle Ages
www.utah.edu/umfa/intro.html

About the Author

Melinda Lilly is the author of several children's books. Some of her past jobs have included editing children's books, teaching pre-school, and working as a reporter for *Time* magazine. She is the author of *Around The World With Food & Spices* also from Rourke.